Mesmerizing
Math
Puzzles

Rodolfo Kurchan

Offical
American Mensa
Puzzle Book

Sterling Publishing Co., Inc.
New York

Library of Congress Cataloging-in-Publication Data

10   9   8   7   6   5   4   3   2   1

Published by Sterling Publishing Company, Inc.
387 Park Avenue South, New York, N.Y. 10016
Originally published in Argentina by United Games SRL
under the title *Diversiones con Numeros y Figuras*
©2000 by Rodolfo Kurchan
©2000 by United Games SRL
English translation © 2001 by Sterling Publishing Company
Distributed in Canada by Sterling Publishing
% Canadian Manda Group, One Atlantic Avenue, Suite 105
Toronto, Ontario, Canada M6K 3E7
Distributed in Great Britain and Europe by Cassell PLC
Wellington House, 125 Strand, London WC2R 0BB, England
Distributed in Australia by Capricorn Link (Australia) Pty Ltd.
P.O. Box 6651, Baulkham Hills, Business Centre, NSW 2153, Australia

Sterling ISBN 0-8069-3709-2

CONTENTS

PREFACE

The simplest numbers, the natural numbers (1, 2, 3, 4, 5, 6, ...) offer the richest and most surprising tool kit of mathematics. This book begins with a collection of new problems with natural numbers. They present challenges that can be solved with ordinary methods. On the one hand, there are the simplest methods: adding, subtracting, multiplying, and dividing, which we learned in elementary school. On the other hand, there is a traditional way to solve puzzles, which consists of simply confronting the problems with the spirit of a player. A player enjoys the challenge of his or her opponent, which leads to testing, probing, and searching for an opponent's weaknesses—and overcoming them.

One chapter is devoted to new problems using traditional geometry. It contains figures that must be cut into equal parts. Solving these cutouts requires methods that are even more ordinary than those of number puzzles. Visual intuition and mental flexibility in manipulating the pieces are what count here. The game of pentominoes presents further challenges. This is one of the world's most popular types of jigsaw puzzle, and Rodolfo Kurchan is a leading expert on the subject. The last chapter contains a collection of miscellaneous problems: dominoes, moves across a board, and so forth.

Mesmerizing Math Puzzles enriches the classical field of mathematics with challenges that are really new and original. In short, it offers raw material for intellectual enjoyment and entertainment.

Rodolfo Kurchan, the inventor of all the book's puzzles, has been solving, discovering, inventing, and collecting ingenious problems and puzzles since adolescence.

His contributions have appeared in numerous publications: *Journal of Recreational Mathematics* (U.S.), *Cubism for Fun* (Holland), *World Game Review* (U.S.), *Games* (U.S.), *Humor Registrado* [Recorded Humor] (Argentina), *El Acertijo* [The Puzzle] (Argentina), etc. Currently, he manages *Puzzle Fun*, a publication devoted to pentominoes. Those interested may consult: http://pagina.de/rkurchan.

His enthusiasm and skill have made him a world traveler. He participates in competitions and international meetings: *World Puzzle Championship* (New York, 1992), *International Puzzle Party #17* (San Francisco, 1997), *IPP #19* (London, 1999), *G4G4* (Fourth meeting in honor of Martin Gardner, Atlanta, 2000). He is a member of the Argentine group *Los Acertijeros* (The Riddle Makers).

Rodolfo Kurchan is an avid athlete, especially enthusiastic about basketball. Along with his father and brother, he manages a coin and stamp collecting business.

Iván Skvarca, another member of *Los Acertijeros,* was responsible for applying art and ingenuity to the text and graphics.

—Jaime Poniachik

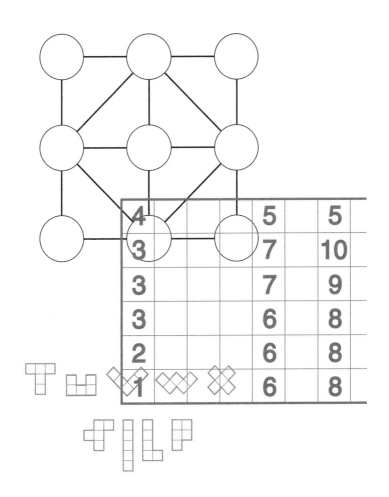

NUMBERS

1. The Hour of the Bat

Even upside down a bat can continue to use its digital watch, because some hours are the same even when viewed in reverse. For example, this is what happens with digital 12:21, which when viewed upside down, is the same.

We call these hours reversible.

What is the shortest interval between two reversible hours?

(Disregard the colon that separates hours from minutes. Time cannot begin with zero.)

2. Volleyball Tournament

There are 163 teams in a volleyball tournament.

The tournament is played using double elimination: the team that loses two games is eliminated from the competition.

How many games will be played in the entire tournament?

3. Magic Posters

A kit has four posters that can be changed magically.

There are three numbers that are whole and greater than zero, that add up to fifteen, according to the instructions of the magician.

"What are the numbers?" you ask.

The magician's instruction booklet says: "These four magic posters will give you the answer. Two of them are correct. The other two are decidedly incorrect."

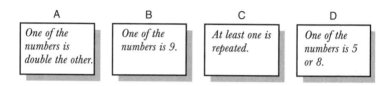

A	B	C	D
One of the numbers is double the other.	*One of the numbers is 9.*	*At least one is repeated.*	*One of the numbers is 5 or 8.*

"What are they, what are they?" you ask. It's going to take me all week to figure out."

Then you concentrate deeply on the problem.

On Monday you believe the right posters are A and B. **What is the answer?**

On Tuesday you are sure the right posters are A and C. **What is the answer?**

On Wednesday you are confident the right posters are A and D. **What is the answer?**

On Thursday you suppose the right posters are B and C. **What is the answer?**

On Friday you think the right posters are B and D. **What is the answer?**

On Saturday you guess that the right posters are C and D. **What is the answer?**

On Sunday you're content, knowing you've got the answers.

Find the six solutions.

4. Lottery in Wonderland

The Queen of Hearts buys a lottery ticket. The number has five digits.

"Bah!" she says. "It's not a palindrome, because of one of the digits."

Disappointed, she gives the ticket to Alice.

"Great!" Alice says. "If I take the number preceding it and multiply each digit by the others in succession, and if I take the number following it and multiply each digit by the others in succession, the difference between the two is 1."

What is the lottery number?

5. Honest Panels

A panel is honest when all numbers that appear inside its boxes are accurate.

To understand the idea, let's begin by building a panel with three boxes cut out.

Each box is labeled underneath with a number. In the beginning each box is empty.

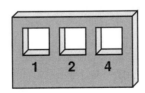

We then place a number inside each box that tells us how many of the numbers on the label below are in the entire panel.

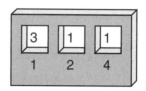

So, we can read it like this: on the panel (counting the numbers underneath the boxes as well as those inside the boxes) are 3 ones, 1 two, and 1 four. You can test it. You'll see it's correct.

Fill in the panels below in the same way. (Each one is separate from the others.)

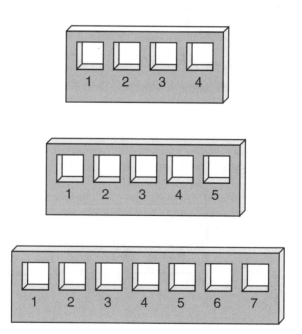

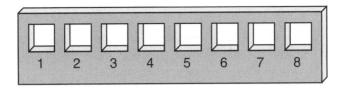

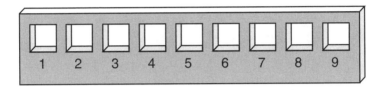

6. Dance of the Primes

At a party there is an odd number of girls and the same number of boys.

Each girl has a number, beginning with 1, and in succession: 2, 3, 4, etc.

The same is true of the boys. Each has a number, beginning with 1, and in succession: 2, 3, 4, etc.

When the music plays, each girl begins to dance with a boy. The only condition in choosing a partner is that the sum of the two numbers—hers plus his—must equal a prime number. (A prime is a number that is divisible only by itself and 1.)

At this time, all the girls and boys are dancing.

Which boy is girl number 1 dancing with?

7. Exact Ending

When dividing 512 by the last digit, 2, the quotient is exact. There is no remainder. When dividing 819 by the last digit, 9, the result is exact.

We say these are numbers with an exact ending.

Find the lowest nine consecutive numbers with exact endings.

(Of course, the numbers from 1 to 9 are one answer. But this is simple. Disregard it.)

8. The Highest with an Exact Ending

In this number, which uses all the numbers from 1 to 9, each group of three consecutive numbers forms a number with an exact ending. Just as with the previous problem, when dividing the number by the last digit, the result is exact.

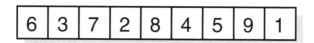

Observe: 637 is divisible by 7; 372 is divisible by 2; 728 is divisible by 8; 284 is divisible by 4; 845 is divisible by 5; 459 is divisible by 9; and 591 is divisible by 1.

What is the highest number using nine digits and no repeats that meets this condition?

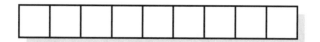

9. The Opponents

"How many soldiers do you see?" asks Julius Caesar. "It's a four-digit number. The first two digits together form a number squared. The two digits in the middle together also form a number squared. And likewise, the last two digits."

"That's nothing," says Marc Antony. "I have a three-digit number of soldiers, which is a number cubed. In addition, it's the product of multiplying the digits of your number of soldiers by each other."

Seeing that they are not making an impression, they draw their swords.

How many soldiers does each have?

10. The Lottery

The small balls are numbered from 1 to 9. The lottery official turns the drum and draws three balls, each with a different number.

"In what order do I put the three balls?" is the official's question.

It can be done in only six different ways. In each case there is a three-digit number that is divisible by a different number, as indicated below.

What were the numbers drawn?

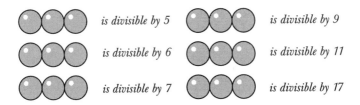

is divisible by 5   is divisible by 9

is divisible by 6   is divisible by 11

is divisible by 7   is divisible by 17

11. Cervantes Mania

Cervantes separates numbers into three groups.

Why?

(He has a reason, unusual but strict, for doing it. When you discover it, you'll also know in which group to put 14.)

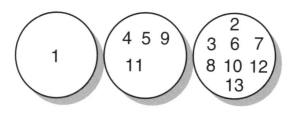

12. Middle Numbers

After the parrot says a number, Robinson multiplies its digits and Crusoe adds them. If the difference between the two results is halfway between the two, they announce that the parrot's number is a middle number, and they celebrate.

For example, if the parrot says 75, Robinson multiplies 7 × 5 and gets 35; meanwhile, Crusoe adds 7 + 5, which makes 12. The difference between 35 and 12 is 23, which is between the two results; therefore, 75 is a middle number.

One afternoon, under a coconut palm, the question is "What is the lowest middle number?"

13. The House of the Triplets

"I'll go with you to your house," the young man says to the triplets while they are walking down the street. "Which one is it?"

"The house number has three digits," one triplet says. "If we reverse it and insert a digit, the new number is triple the old number."

What is the house number?

SEQUENCES

The numbers or figures of each sequence follow strict and rigorous criteria, although not always apparent. Test your mental strength by finding the secret to each sequence. If you feel lost, the titles may help you.

14. In the Verses

What's the secret to this sequence? What is the next number?

1, 7, 11, 27, 77, 117, 127, ?

15. Even So

What's the secret to this sequence? What is the next number?

4, 5, 9, 11, 12, 13, 14, ?

16. Absent

What's the secret to this sequence? What is the next number?

2, 4, 6, 30, 32, 34, 36, 40, 42, 44, 46, 50, 52, 54, 56, 60, 62, 64, 66, 2000, ?

17. Ease and Comfort

What's the secret to this sequence? What is the next number?

1, 3, 11, 17, 111, 117, 317, ?

18. Close Encounter

What's the secret to this sequence? What is the next number?

1, 9, 19, 99, 119, 199, 999, ?

19. Inversion

The numbers from 1 to 10 are arranged according to strict and unusual criteria. What's the secret to this sequence?

3, 9, 1, 5, 10, 7, 2, 4, 8, 6 ?

20. The Calculator

What's the secret to this sequence? What is the next number?

 ?

21. Double Turn

What's the secret to this sequence? What is the next number?

1, 2, 4, 8, 61, 221, 244, ?

22. The End

What's the secret to this sequence? What is the next number?

7, 10, 11, 13, 14, 15, 16, 17, 18, 19, 27, ?

CODED SUMS

Each coded sum hides a numerical operation.

Replace each letter with a number so that the correct sum or multiplication appears.

Each letter must be replaced by a number. When a letter is repeated, the corresponding number must also be repeated. A zero cannot be in the leftmost position. In some cases, a special condition must be kept in mind.

23. Water Everywhere

```
   W  A  T  E  R
   W  A  T  E  R
   W  A  T  E  R
+  W  A  T  E  R
───────────────
   O  C  E  A  N
```

24. Fun and Games

```
   C  A  R  D
+  C  A  R  D
──────────────
G  A  M  E  S
```

Do not use a zero for this one.

25. Even Steven

```
    E  V  E  N
    E  V  E  N
+ N  E  V  E  R
─────────────────
  P  R  I  M  E
```

Use only numbers between 1 and 8 for this one.

26. Reading

```
    P  A  G  E
    P  A  G  E
    P  A  G  E
    P  A  G  E
    P  A  G  E
+ P  A  G  E
─────────────
  B  O  O  K
```

Do not use a zero for this one.

27. Economics

```
    C  O  I  N
    C  O  I  N
+  C  O  I  N
─────────────
 M  O  N  E  Y
```

28. Washing Up

```
  S  O  A  P
×        6
─────────────
  B  A  T  H
```

Do not use the zero for this one.

29. Gamesmanship

```
  R  O  O  K
+ R  O  O  K
-----------
C  H  E  S  S
```

30. Time

```
  D  A  Y  S
×         7
-----------
W  E  E  K
```

Do not use the nine for this one.

31. Fruit

```
  A  P  P  L  E
  A  P  P  L  E
+ A  P  P  L  E
--------------
  F  R  U  I  T
```

32. Affection

```
  L  O  V  E
  L  O  V  E
+ L  O  V  E
-----------
H  E  A  R  T
```

Do not use the zero for this one.

33. Romance

```
  K  I  S  S
  K  I  S  S
  K  I  S  S
+ K  I  S  S
-----------
  L  O  V  E
```

Do not use the zero for this one.

34. Happy

```
  S  M  I  L  E
  S  M  I  L  E
+ S  M  I  L  E
--------------
  H  A  P  P  Y
```

Do not use the zero for this one.

35. Crybaby

```
  C  R  Y
  C  R  Y
+ C  R  Y
--------
  B  A  B  Y
```

Do not use the zero for this one.

36. Calculation One

In its debut, the "Calculating Horse" places the numbers from 1 to 9 in the boxes, using each number only once, and the result is a correct numerical operation.

How does he do it?

☐ × ☐ = ☐ × ☐ = ☐ + ☐ + ☐ + ☐ + ☐

1 2 3 4 5 6 7 8 9

37. Calculation Two

In the following exercise, the Calculating Horse places the numbers from 1 to 9 in the boxes and, using each number only once, again achieves a correct numerical operation.

How does he do it?

(Keep in mind that you must first do the multiplication and then the addition.)

☐☐ × ☐ = ☐☐ × ☐ = ☐ × ☐ + ☐

1 2 3 4 5 6 7 8 9

38. Calculation Three

In the last exercise, the Calculating Horse places the numbers from 1 to 9 in the boxes, and, using the numbers only once, again manages to achieve a correct numerical operation.

Can you do it too?

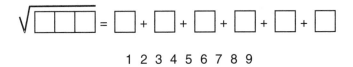

1 2 3 4 5 6 7 8 9

39. The Witch's Charm

A witch needs a new charm. She leaves the forest, draws a diagram like the one below on the bark of a birch tree, and then places the numbers 1 to 9 in the empty circles. Using each number only once, she makes the four small squares, the corners of the large square, and the diamond add up to the same number.

What is the charm?

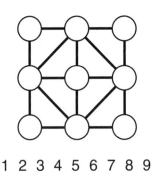

1 2 3 4 5 6 7 8 9

40. Good Neighbors

In order to move out of the neighborhood, the numbers 1 through 9 must meet two conditions.

One, boxes with two consecutive numbers may not touch at the sides or diagonally.

Two, boxes with two numbers that can be divided by each other may not touch at the sides or diagonally. (The exception is 1, because otherwise it would be impossible to place it.)

What does the new neighborhood look like?

The first attempt by the real estate agent fails because 8 and 9 are consecutive numbers and their boxes are adjacent. In addition, 4 is divisible by 8 and the two are neighbors; the same is true of 2 and 8.

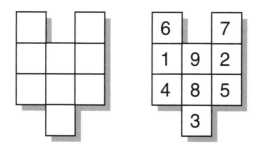

41. National Security

The Pentagon is on maximum alert.

"Our enemies are threatening," reports the secretary to the chairman of the joint chiefs, "to use the numbers 1 through 9—each number only once—to form various numbers squared."

A hush descends upon the room.

"If you add them," asks the chairman, "what is the lowest possible sum?"

Help them, reader. Using all the numbers from 1 to 9, form numbers squared.

What is the lowest possible sum that can be obtained?

For example, if you form the numbers 15376 (which is 124^2), 289 (which is 17^2), and 4 (which is 2^2), the sum is 15669. There is an even lower sum.

42. Strongbox One

The Masked Thief does not always use picklocks and explosives to do the job.

In order to open this box, he must place the numbers 1 to 9 in the empty squares, so that all arithmetical operations are correct.

How does he open it?

(Keep in mind that at the bottom right, where there are two adjoining empty squares, he must place a number in each box, forming a two-digit number. That number will be the end result of the other operations.)

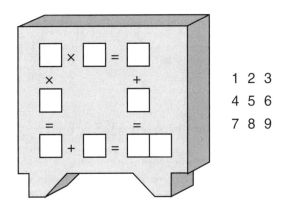

43. Strongbox Two

Box Two also opens by placing the numbers 1 to 9 in the empty squares and also forms correct arithmetical operations.

How does the Masked Thief do it?

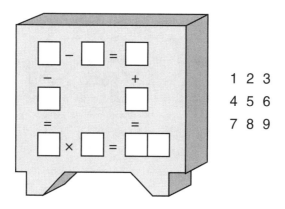

44. Strongbox Three

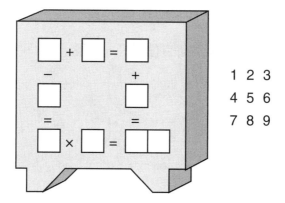

Box Three opens the same way.

How?

45. Strongbox Four

How does he open Box Four?

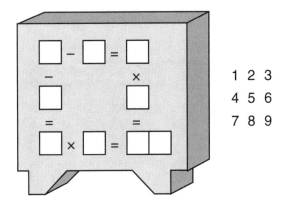

1 2 3
4 5 6
7 8 9

46. Strongbox Five

Box Five is the same as the others, but it also has a dangerous diagonal combination.

Can he open it this time?

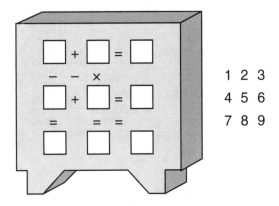

1 2 3
4 5 6
7 8 9

47. Strongbox Six

To open Box Six, the Masked Thief must place eight of the numbers from 1 to 9 in the empty squares.

Just as before, he may use the numbers only once, and each square must have only one number.

The operations must be read in the direction heading toward the equal signs. Take the third row, for example: the number on the right is divided by the number in the middle, and the result is in the box on the left.

How does the strongbox open?

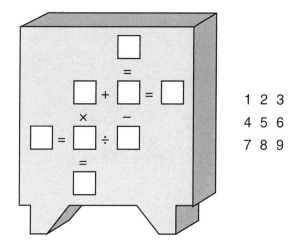

1 2 3
4 5 6
7 8 9

48. Strongbox Seven

Here also the two adjoining squares indicate the formation of a two-digit number, and the operations must be read in the direction heading toward the equal sign.

How is it opened?

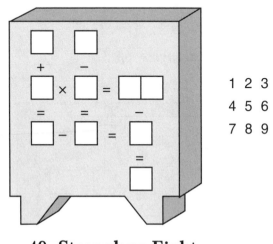

1 2 3
4 5 6
7 8 9

49. Strongbox Eight

This is a new strongbox.

How does it open?

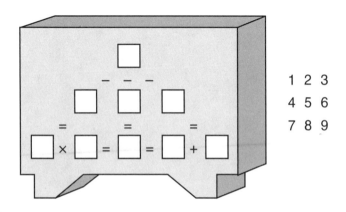

1 2 3
4 5 6
7 8 9

50. Strongbox Nine

This is the last strongbox. It opens just like the others.

How?

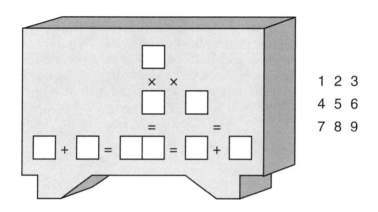

1 2 3
4 5 6
7 8 9

51. Impossible Love

The lovers are hugging each other and crying.

"Our parents are demanding that we distribute the numbers from 0 to 9," whispers Juliet.

"Let's try not to separate ourselves too much," Romeo proposes. "With these numbers, let's form two numbers that are as close together as possible—97531 and 86420, for example. The difference is only 11111."

"There must be something better," Juliet says.

A skilled swordsman interrupts this idyll, but we're not interested now. Help the young lovers, reader.

Find the two closest numbers formed by the numbers 0 to 9. Use each number only once.

52. The Spy in Trouble

"The key, I need the key!"

The person shouting is James Bond. The young woman looks at him.

"It was a number made up of eight different digits," she says.

"Yes, that's right," Bond says. "The first two digits form a number that is half the number formed by the third and fourth digits. The fourth is one-third the number formed by the fifth and sixth digits, and the sixth is one-fourth of the number formed by the seventh and eighth digits. But what is that wretched number?"

"You still haven't found it," she says, examining the polish on her fingernails.

What is the number?

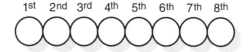

1st 2nd 3rd 4th 5th 6th 7th 8th

53. The Conquest of Space

On his first voyage, the astronaut was in orbit from 6:34 on May 8 until 1:02 on July 9, a total of 61 days, 19 hours, and 28 minutes.

"How strange," he said upon returning to Earth. "The first date is written 5/8, 6:34, and the second, 7/9, 1:02. The two use all numbers between 0 and 9, each only once."

The second trip had the same outrageous characteristic: the departure and return dates, both during the same year, were written using all the numbers 0 to 9 only once. But this trip was the longest that could have been made under that condition. Times can start with 0, so "0:13" is 12:13 a.m.

On what date and at what time was the takeoff, and on what date and at what time was the return?

The astronaut's third voyage had the same characteristics, but it was the shortest trip possible.

On what date and at what time was the takeoff, and on what date and at what time was the return?

54. The Teacher's Voice

The teacher says to the pupil:

"When you can place the numbers 1 through 9 in the empty squares of this diagram, so all the lines, both diagonally and vertically, add up to the numbers shown, then you will be smart."

Are you smart, reader?

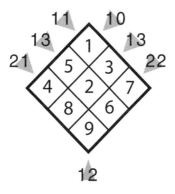

FIGURES TO BE DIVIDED

Most of the drawings in this chapter must be divided into *exactly equal* parts. The number of parts is indicated in each problem. The parts may be rotated or turned, like a reflection in a mirror. The cutout does not necessarily have to follow the lines of the grid, which serve only to lend shape to the figures. Disregard small decorative accessories such as eyes and black spaces.

55. The Elephant

Divide the elephant's head into *two* equal parts. Disregard the eye and the tusk.

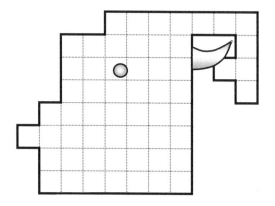

56. The Board

Divide the board into *two* equal parts. The black boxes are a hole. Disregard them.

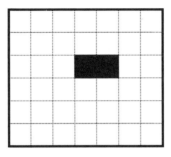

57. The Hammer

Divide the hammer into *two* equal parts.

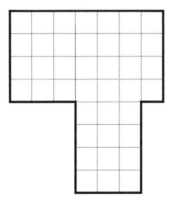

58. Another Board

The corners of this board have been cut out. Divide it into *two* equal parts.

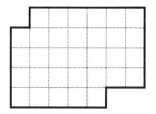

59. The Factory

Divide the factory into *two* equal parts. Disregard the plume of smoke.

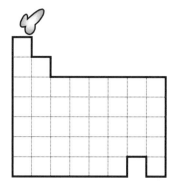

60. The Sumo Wrestler

Divide this sumo wrestler into *two* equal parts. Don't be intimidated by the eyes and the mouth.

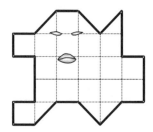

61. The Duck

Divide the duck into *four* equal parts. Disregard the eye.

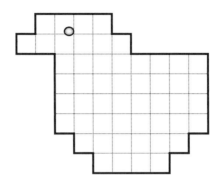

62. The Marathon Runner

Divide the marathon runner into *four* equal parts. The shoes don't count, and you can ignore the black box, which is a hole.

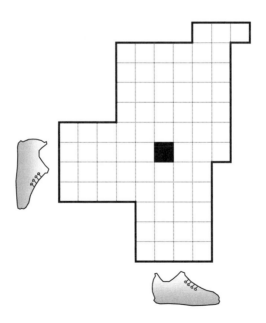

63. The Faucet

Divide the faucet into *four* equal parts. Don't let the drop of water spread.

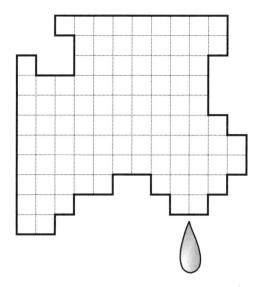

64. The Mask

Divide the mask into *five* equal parts. The black box is a hole. Disregard it.

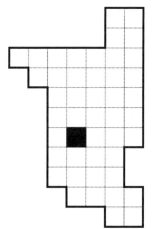

The Building

In the five problems that follow, the silhouette of the same building must be divided. In each, some windows are closed. These are designated by black boxes. They are not included in the figure to be divided. As always, the number of parts changes from one problem to the next, and the parts must be equal in shape and size. The cutouts do not necessarily have to follow the lines of the grid, which are only indicators of the shape.

65. Building in Three

Divide the building below into *three* equal parts.

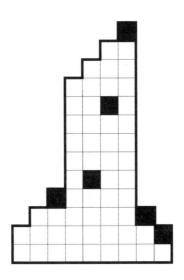

66. Building in Four

Divide the building below into *four* equal parts.

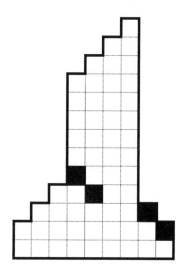

67. Building in Five

Divide the building below into *five* equal parts.

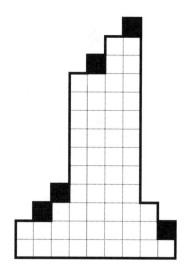

68. Building in Six

Divide the building below into *six* equal parts.

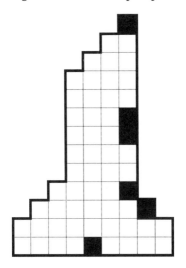

69. Building in Seven

Divide the building below into *seven* equal parts.

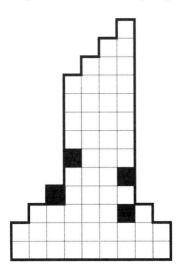

70. The Inheritance

When the family patriarch decided to divide the land among his four heirs, he wisely divided it into *four* equal parts and made sure each part had no houses with different letters.

How was the land divided?

Cut only along the lines.

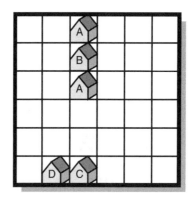

71. Division of AB

Divide figure AB into the fewest number of pieces that will subsequently fit into a square with five smaller squares per side.

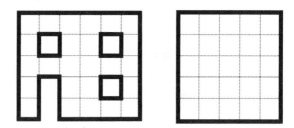

72. Division of ABC

Divide figure ABC into the fewest number of pieces that will subsequently fit with ease into a square with six smaller squares per side.

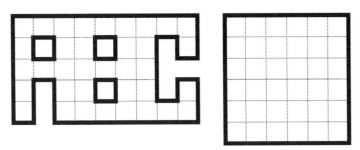

73. Family Inheritance

The family grows. Cut the rectangular tablecloth measuring 6 × 9, which was inherited from Grandmother, into the fewest possible number of pieces. With these pieces, simultaneously fill four tablecloths of 2 × 2, 3 × 3, 4 × 4, and 5 × 5 squares per side.

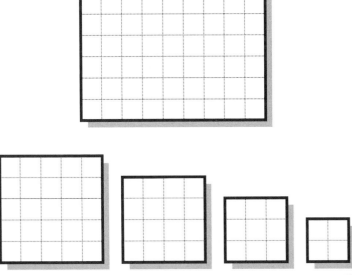

PENTOMINOES

Take five equal squares and join them end to end. There are twelve different ways of doing it, leaving aside those shapes that are the same because of rotation or symmetry. These twelve figures are called pentominoes. Solomon Golomb was the first to explore these figures systematically, in the 1950s, and he was the one who named them.

Placing them as shown below, they look like letters, and thus we can more easily name them. As a mnemonic device you may think of the last seven letters of the alphabet and of the word *Filipino* (FILPN).

Unless otherwise instructed, the pieces may be rotated and mirror themselves—that is, front and back are the same.

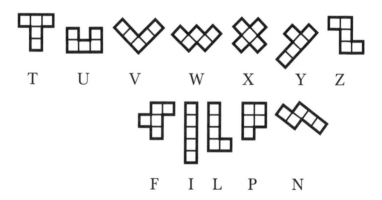

T   U   V   W   X   Y   Z

F   I   L   P   N

74. No Touching

This grid is covered with the twelve pentominoes. Their sides and corners do not touch.

If the black boxes do not contain pentominoes, how must the grid be covered?

(Some of the white squares will remain uncovered.)

75. A Patio with Tiles

The patio below was covered with the twelve pentominoes.

When two or more squares contain the same number, it means that those squares are under the same piece.

How is the patio covered?

4				5		5		12	
3				7		10		10	
3				7		9		12	
3				6		8		9	
2				6		8		9	
1				6		8		11	

76. Another Patio with Tiles

This patio was also covered with the twelve pentominoes and each point is under a different piece.

How is the patio covered?

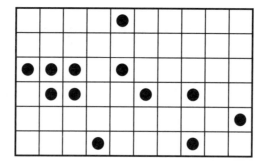

77. The Kitchen Floor

My kitchen floor is an 8 × 8 grid with nine black tiles. I want to completely cover the white tiles with nine equal pentominoes.

Which pentomino must I use, and where do I place it?

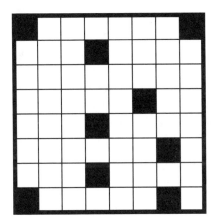

78. Reconstruction 1

The vertical lines indicate the borders of the pieces.

Where must the 12 pentominoes be placed in order to cover the rectangle?

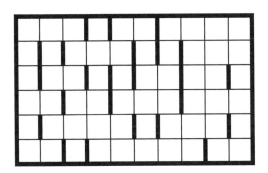

79. Reconstruction 2

The value shown in the illustration was assigned to each pentomino.

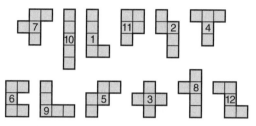

How must the 12 pentominoes be placed so that the numbers underneath the board on the following page represent the sum of the values of the pentominoes in each column?

(Each pentomino is added once per column, but, if it is part of two or more columns, it is counted in each one of the columns in which it appears.)

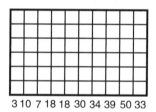

3 10 7 18 18 30 34 39 50 33

80. Reconstruction 3

The thick lines indicate the borders of the pieces.

How must the 12 pentominoes be placed in order to cover the rectangle?

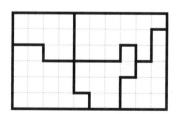

81. A Game of Dominoes

Melchior, Gaspar, Balthazar, and Roy, seated in that order, were playing a game of dominoes.

The game has 28 dominoes, one for each combination of the numbers 0 to 6. In the traditional game, players must place matching combinations next to each other.

Melchior began with domino 6·4.

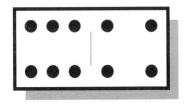

After placing the tenth piece on the table, the 3·6 next to the 0·3, Gaspar ended the game. No one could make another move.

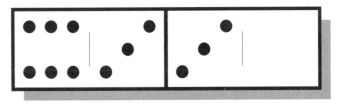

The total number of pips of all Melchior's dominoes was 29; Gaspar's, 23; Balthazar's, 20; and Roy's, 6.

Determine which dominoes were played and in what order.

82. Hopscotch of Prime Numbers

The tiles are numbered from 1 to 100 and by placing them in order, we can construct a patio. The patio may be as wide or as narrow as we want. It doesn't matter if the last row of tiles is incomplete.

Once we've constructed it, we want to go from the first row to the last by jumping to tiles with prime numbers. If we want to move from one tile to another, the two must be touching on the side or diagonally.

Remember, a prime number is one that has only two divisors: the number itself and 1. (A table that appears at the end of this book lists all prime numbers up to 1013.)

With a patio three tiles (that is, three columns) wide, it is impossible to reach the last row. One can only reach 23. Neither 25, 26, nor 27 is a prime number.

1	2	3
4	5	6
7	8	9
10	11	12
13	14	15
16	17	18
19	20	21
22	23	24
25	26	27
28	29	30

If the patio has 94 columns, it's possible to reach the last row, but such a patio is exceptionally wide.

1	2	3	4	5	6	7	8	9		91	92	93	94
95	96	97	98	99	100								

What is the narrowest patio where it's possible to move from the first row to the last by jumping only on tiles with prime numbers?

83. Mancala

Mancala is a traditional African game. The game begins with piles of stones arranged in rows. A move consists of picking up an entire pile of stones, then moving along the row, leaving a stone on each pile visited. The move may be toward the right or left. Changing directions after beginning is not permitted before reaching the end of the row. After that, if there are still stones remaining in the hand, the player may turn in the other direction, leaving one stone on each pile. One stone is placed on the farthest pile.

For example, let's begin with three piles, with one, two, and three stones respectively.

How would you achieve piles of three stones, two stones, and one stone?

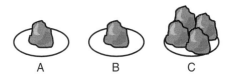

Three moves are sufficient.

First, the stones of pile B are distributed toward the right, and are as follows:

Then, the stone of pile B is distributed to the left, as follows:

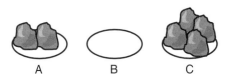

Finally, the stones of pile C are distributed. (Since the pile is on the extreme right, the stones are always distributed toward the left.)

A   B   C

Here are five other challenges. The numbers indicate how many stones are in each pile, beginning with pile A. In order to record the moves, write down which pile the stones are being taken from and which side the stones are moving to.

Challenge 1
Move from 1·2·3·4 to 4·3·2·1.

Challenge 2
Move from 1·2·3·4·5 to 5·4·3·2·1.

Challenge 3
Move from 0·1·2·3·4 to 2·2·2·2·2.

Challenge 4
Move from 1·2·3·4·5 to 3·3·3·3·3.

Challenge 5
Move from 1·2·3·4·5·6 to 6·5·4·3·2·1.

84. The Good Investor

To play this game you need seven chips numbered 1 to 7.

Arrange them in their starting position and move them to the final position in the fewest number of moves.

starting position

final position

A move consists of taking three consecutively numbered chips and inverting them. For example, begin like this.

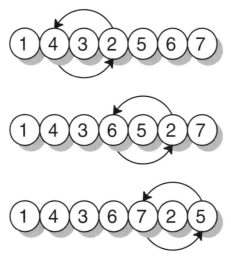

How many moves does the solution require?

85. The Eccentric Exam

A teacher in a pass/fail class gives a very strange exam. Each correct response is worth one point, and each incorrect response counts as minus half a point. It is not necessary to answer every question. No points are deducted for unanswered questions. A score of 5 is needed to pass. There are ten questions in the test. After thinking a moment, his students did not answer six or seven questions, but they did answer five or eight.

Why?

86. Unaligned Pawns

What is the highest number of pawns that can be placed on a board so that no three are aligned horizontally, vertically, or diagonally?

Pawns must be placed in the boxes, forming a group we call "connected"—that is, there can be no isolated pawns or groups of pawns. Each pawn must be a neighbor to at least one pawn, either vertically or horizontally, and one must be able to move from one pawn to another by a path that crosses only boxes with pawns.

The answer is *four*. Arranging the five pawns any other way results in the alignment of three of them.

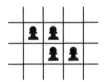

What is the highest number of pawns that can be placed on a board so that no four are aligned?

(Remember, the pawns may not be isolated, and the lines to avoid are verticals, horizontals, and all types of diagonal. Of course, the pawns are always in the center of the boxes. There are no tricks.)

87. Square Stamps

We have rubber stamps that are squares. Here we see a complete set of stamps in sizes 1, 2, 3, and 4.

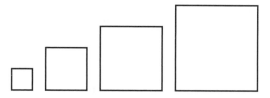

Using each stamp only once we can make seven small squares of size 1.

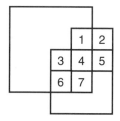

How many small squares in size 1 can be formed with five stamps in sizes 1, 2, 3, 4, and 5? And how many with six and seven stamps?

Within the borders of the large stamp, stamps 1, 2, 3, and 4 can form four size-1 squares.

How many squares of size 1 can be formed with stamps of sizes 1, 2, 3, 4, and 5? And how many with six and seven stamps?

88. The Ghost Detective

A detective crossed a rectangular field. In the first box he wrote "1." Later he passed a neighboring box, horizontally or vertically but not diagonally, and there he wrote "2." In this manner he crossed all twenty boxes, marking successive numbers in each one until he reached the last box, in which he wrote "20."

The sum of the three marked boxes is 12.

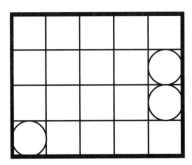

Reconstruct the detective's path.

89. The Detective and His Double

Two detectives crossed over this field.

Each began his trip with number 1 and ended with 10. Neither detective visited a box visited by the other detective.

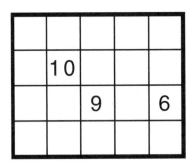

What route did each one take?

90. Banana Peel

When a checker moves like a banana peel, it moves horizontally or vertically and only stops when it reaches the edge of the board or when it runs into another checker.

On this board with five boxes per side, the objective is to move any checker to the central box using the fewest number of banana-peel moves.

The best-known solution uses ten moves.

Do it or improve on it.

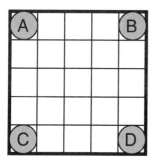 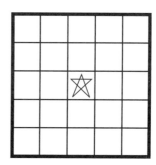

91. Fifteen Sweep

This is a traditional game using a Spanish deck of cards. It has four suits of ten cards each, for a total of 40 cards. The values of these cards ranges from 1 to 10.

Three cards are dealt to each player. Moving around a circle, each player in turn places one card on the table. When two or more cards on the table add up to 15, the player who put down the last card (that is, the one that reaches 15) picks up and holds those cards.

These rules are the only ones you need to know in order to answer the following questions about Fifteen Sweep.

1. What is the minimum number of cards on the table that will force a player to pick up the cards, regardless of the card played?

("Forced to pick up cards" means that some combination has reached 15.)

2. What is the highest number of odd-numbered cards that could be on the table?

3. What is the highest number of cards that could be on the table and still have an equal number of even-numbered and odd-numbered cards?

4. When the hand is over, each player counts the number of cards he or she has picked up. How many cards could never be the number picked up?

92. Iron Way

To service a route between A and B, there are eight sections of rail of successive lengths from 1 to 8. The route passes through three tunnels.

Each section must begin and end on another section or in one of two cities. All sections must be used.

How is the route constructed?

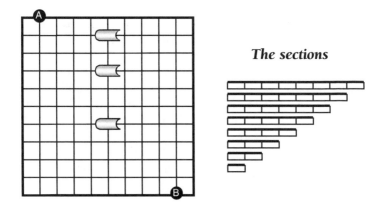

The sections

93. The E-Mail Riddle

Mr. Spock and Captain Kirk were playing tic-tac-toe by e-mail. Each in turn sent his move by e-mail until Spock won the game.

As a radiation cloud passed overhead, the order of the messages became scrambled.

In what order were they sent?

A. I will place an X, completing the other diagonal.

B. There is an empty box to the right of your last move, but I am not going to move there; I will instead place an O in the box below your X, which is also empty.

C. I will place one O in the grid, one box to the right and two boxes below your last X (or to the left of my last O, which is the same thing).

D. I will place my O, completing the first column on the left.

E. I will place an X in the center box.

F. What happened? I lost!

G. I will place an X to complete the horizontal line.

H. I am moving on the same diagonal as your last move.

SOLUTIONS

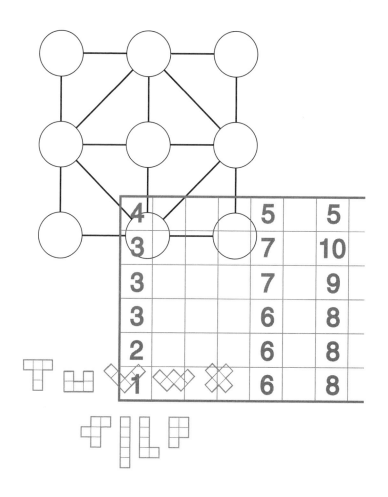

NUMBERS

1. The Hour of the Bat

The shortest interval between two reversible hours is between 9:56 and 10:01 and is five minutes.

9:56 10:01

2. Volleyball Tournament

There will be a total of 324 or 325 games, depending on whether the champion loses one game or none. Since there are 163 teams and only one will win, the other 162 must be eliminated. In order to be eliminated, each team must lose two games, hence the total of 324 games.

3. Magic Posters

Monday's solution is 2, 4, and 9.
Tuesday's solution is 3, 6, and 6.
Wednesday's solution is 3, 4, and 8.
Thursday's solution is 3, 3, and 9.
Friday's solution is 1, 5, and 9.
At last, Saturday's solution is 5, 5, and 5.

4. Lottery in Wonderland

The lottery ticket number is 11110.

This is not a palindrome because of one number, and if we multiply the digits of the number that precedes it (11109), the result is 0, whereas the result of multiplying the digits of the next number is 1 (11111).

5. Honest Panels

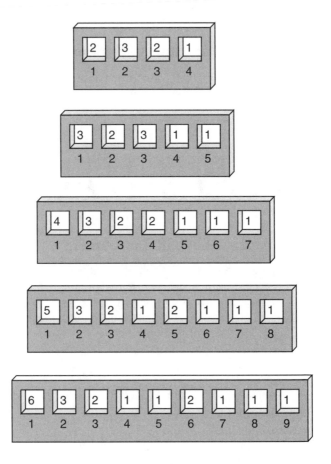

There is no solution to a panel labeled 1 to 6.

6. Dance of the Primes

Girl number 1 is dancing with boy number 1. There are an odd number of girls, and odd numbers exceed evens by one. In order for the sum of each couple to equal a prime number, the girls with odd numbers must dance with boys with even numbers, because the sum of two odd numbers is a multiple of 2 and a multiple of 2 cannot be a prime number. The only multiple of 2 that is a prime number is 2 itself. Therefore, the only odd number, which when added to another number equals 2, is 1.

7. Exact Ending

The numbers between 2521 and 2529 yield an exact result: $2521 \div 1 = 2521$, $2522 \div 2 = 1261$, $2523 \div 3 = 841$, $2524 \div 4 = 631$, $2525 \div 5 = 505$, $2526 \div 6 = 421$, $2527 \div 7 = 361$, $2528 \div 8 = 316$, and $2529 \div 9 = 281$. It is the lowest possible solution, because 2520 is the lowest number that at the same time can be divided by numbers 1 to 9.

8. The Highest with an Exact Ending

9	8	7	6	5	2	4	3	1

9. The Opponents

Caesar has 1649 soldiers. Marc Antony, 216 soldiers.

10. The Lottery

The numbers drawn are 4, 5, and 9. Thus, 495 is divisible by 5; 954 is divisible by 6; 945 is divisible by 7; 549 is divisible by 9; 594 is divisible by 11; and 459 is divisible by 17.

11. Cervantes Mania

The circle to the left has numbers that have more vowels than consonants. The circle to the right has numbers with more consonants than vowels. Fourteen belongs in the middle circle, the one with numbers that have an equal number of vowels and consonants.

12. Middle Numbers

The lowest middle number is 37.

Multiplying $3 \times 7 = 21$ and its sum is $3 + 7 = 10$. The difference between the two results is 11 $(21 - 10)$. The difference, 11, is less than 21 and greater than 10.

13. The House of the Triplets

The triplet lives in number 351. And if we reverse it (153) and insert 0 (1053), the result is the same as 3×351.

María Eugenia Juiz was co-creator of this problem.

SEQUENCES

14. In the Verses

The number of syllables of each number increases. One has one syllable, seven has two syllables, eleven has three syllables, and so forth. The next number in the series after 127 is 177.

15. Even So

Each of these numbers has an even number of letters. The next number in the series is 18.

16. Absent

None of these numbers has the letter "E" in it. The next number in the series is 2002.

17. Ease and Comfort

Each number has one more "E" than the previous one. The next number in the series, with eight "E's," is 1317 ("one thousand three hundred seventeen"). If you read 1117 as "eleven hundred seventeen," this can also be the answer.

18. Close Encounter

In the numbers in this series, the number of times the letter "N" appears steadily increases by one. The next number in the series is 1199.

19. Inversion

They're arranged alphabetically when read backward.

20. The Calculator

Each number has one more segment than the previous one. Number 8 follows.

21. Double Turn

Each number is double the previous one, but written in reverse. 884 follows (doubling 244 is 488 and then written in reverse is 884), 8671 follows, etc.

22. The End

Each number ends with the letter "n." The next number is 37.

CODED SUMS

23. Water Everywhere

```
   18495
   18495
   18495
 + 18495
   73980
```

24. Fun and Games

```
   6239
 + 6239
  12478
```

25. Even Steven

```
    7872
    7872
 + 27873
   43617
```

26. Reading

```
   1647
   1647
   1647
   1647
   1647
 + 1647
   9882
```

27. Economics

```
    5783
    5783
 +  5783
 -------
   17349
```

28. Washing Up

```
    1243
 ×     6
 -------
    7458
```

29. Gamesmanship

```
    6772
 +  6772
 -------
   13544
```

30. Time

```
    1048
 ×     7
 -------
    7336
```

31. Fruit

```
   18809
   18809
 + 18809
 -------
   56427
```

32. Affection

```
    9538
    9538
 +  9538
   28614
```

33. Romance

```
    2144
    2144
    2144
 +  2144
    8576
```

34. Happy

```
   18743
   18743
 + 18743
   56229
```

35. Crybaby

```
     875
     875
 +   875
    2625
```

DIGITS

36. Calculation One

$6 \times 4 = 8 \times 3 = 1 + 2 + 5 + 7 + 9$

37. Calculation Two

$68 \times 1 = 34 \times 2 = 7 \times 9 + 5$

38. Calculation Three

$729 = 1 + 3 + 4 + 5 + 6 + 8$

39. The Witch's Charm

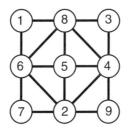

40. Good Neighbors

There are two answers:

8		6
3	1	4
5	7	9
	2	

6		8
4	1	3
9	7	5
	2	

41. National Security

The sum of adding the numbers squared 1, 9, 25, 36, and 784 is 855.

42. Strongbox One

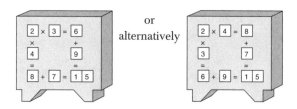

or
alternatively

43. Strongbox Two

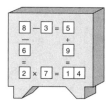

44. Strongbox Three

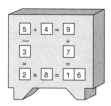

45. Strongbox Four

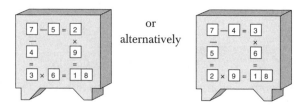

or
alternatively

46. Strongbox Five

47. Strongbox Six

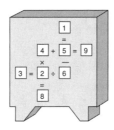

48. Strongbox Seven

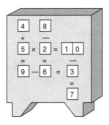

49. Strongbox Eight

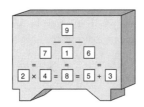

50. Strongbox Nine

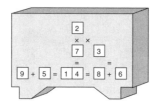

51. Impossible Love

If one forms 50123 and the other 49876, the difference is just 247.

52. The Spy in Trouble

53. The Conquest of Space
The second space voyage, the longest possible, was between January 2 at 0:34 and September 8 at 7:56 (which should be read 0:34 of 1/2 and 7:56 of 9/8). The voyage lasted 250 days, 7 hours, and 22 minutes, in a leap year.

The third space voyage, the shortest possible, was between February 9 at 8:57 and March 1 of 0:46 (which should be read as 8:57 of 2/9 and 0:46 of 3/1). The voyage lasted 19 days, 15 hours, and 49 minutes, in a non-leap year.

54. The Teacher's Voice

FIGURES TO BE DIVIDED

55. The Elephant

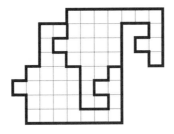

56. The Board

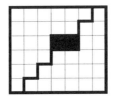

57. The Hammer

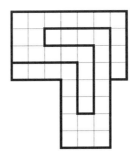

58. Another Board

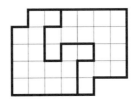

59. The Factory

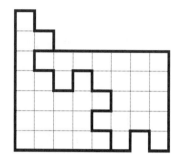

60. The Sumo Wrestler

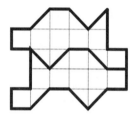

61. The Duck

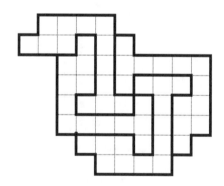

62. The Marathon Runner

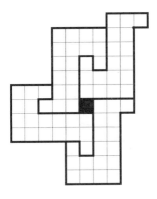

63. The Faucet

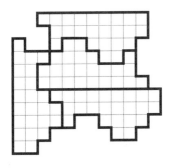

64. The Mask

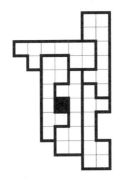

65. Building in Three

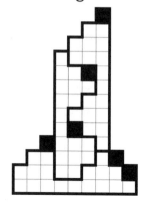

66. Building in Four

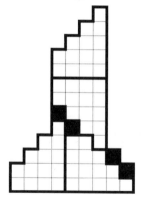

67. Building in Five

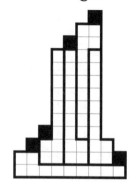

68. Building in Six

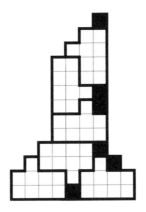

69. Building in Seven

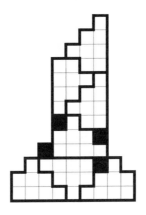

70. The Inheritance

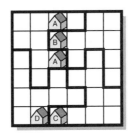

71. Division of AB

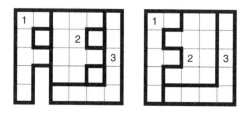

72. Division of ABC

One of the possible cutouts in 5.

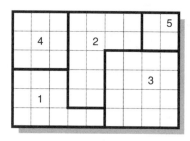 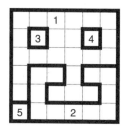

73. Family Inheritance

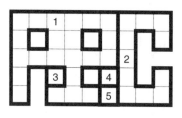

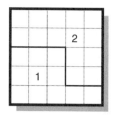

PENTOMINOES

74. No Touching

Note that some pentominoes may occupy slightly different positions (for example, by turning the L around).

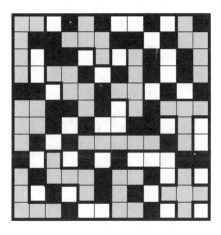

75. A Patio with Tiles

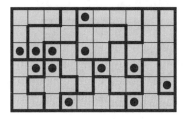

76. Another Patio with Tiles

77. The Kitchen Floor

78. Reconstruction 1

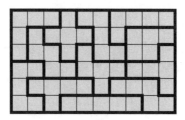

79. Reconstruction 2

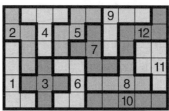

3 10 7 18 18 30 34 39 50 33

80. Reconstruction 3

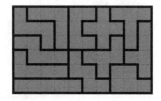

MISCELLANEOUS

81. A Game of Dominoes

For the game to be over, seven dominoes on which the same number appears had to have been played. One of them will be a double, and the other six will have six different numbers. Therefore, in order to be playable, the other three dominoes (we know ten in all were played) had to have six different numbers.

Since the 6 was played twice (in the first and last moves) and we know the scores of each of the players, we can deduce that the number repeated was 6. Therefore, the seven dominoes from 0·6 to 6·6 had to be in the game.

The six numbers remaining range from 0 to 5, and one of the dominoes, as we are told, was 0·3.

The total of Roy's dominoes was six points, and this can only be achieved with 0·3 and 1·2.

Therefore, the third domino that has no 6 among its numbers is 4·5.

Gaspar had 3·6, and we are told his score was 23. We only need to account for the 14 remaining points. We can only arrive at this number with 6·0 and 2·6.

Knowing this, Balthazar had only one way of achieving 20 points: with 4·5 and 6·5.

Finally, we know that Melchior had 6·4 and his total score was 29 points. Therefore, his other two dominoes were 6·1 and 6·6.

With this we know which dominoes were in play, and who had each one. Let's now see how the game evolved.

Melchior played 6·4.

Gaspar placed a domino next to the 6, but we don't know if it was the 6·0 or the 6·2.

Balthazar played the 4·5.

Roy could have played 0·3 or 1·2, but we know that after this move Melchior played the 6·1, so we can deduce that Gaspar played the 6·2; Roy, the 2·1; and Melchior, the 6·1.

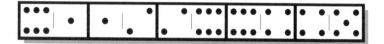

Gaspar played the 6·0; Balthazar, the 6·5; Roy, the 0·3; Melchior, the 6·6. Finally, Gaspar played the 6·3 and ended the match.

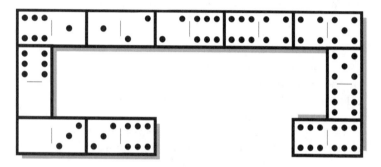

82. Hopscotch of Prime Numbers

The narrowest patio that allows us to go from one point to another by touching tiles with prime numbers is 15 columns wide.

1	2	3	4	5	6	7	8	9	10	11	12	13	14	15
16	17	18	19	20	21	22	23	24	25	26	27	28	29	30
31	32	33	34	35	36	37	38	39	40	41	42	43	44	45
46	47	48	49	50	51	52	53	54	55	56	57	58	59	60
61	62	63	64	65	66	67	68	69	70	71	72	73	74	75
76	77	78	79	80	81	82	83	84	85	86	87	88	89	90
91	92	93	94	95	96	97	98	99	100					

83. Mancala

Challenge 1 can be accomplished in four moves: D to the left, B to the left, D to the left, and C to the left.

Challenge 2 can be accomplished in six moves: C to the right, D to the right, B to the left, C to the right, D to the left, and E to the left.

Challenge 3 can be accomplished in five moves: E to the left, C to the right, B to the right, C to the right, and D to the left.

Challenge 4 can be accomplished in six moves: C to the right, E to the left, B to the right, E to the left, C to the right, and D to the right.

Challenge 5 can be accomplished in nine moves: D to the right, D to the right, F to the left, B to the left, B to the right, C to the left, F to the left, D to the left, and E to the left. (Héctor San Segundo provided this solution.)

84. The Good Investor

One of the possible solutions is accomplished in 9 moves.
1. Invert 567, resulting in 1234765.
2. Invert 347, resulting in 1274365.
3. Invert 127, resulting in 7214365.
4. Invert 436, resulting in 7216345.
5. Invert 216, resulting in 7612345.
6. Invert 345, resulting in 7612543.
7. Invert 125, resulting in 7652143.
8. Invert 214, resulting in 7654123.
9. Invert 123, resulting in 7654321.

Solitaire can be played with more chips, and moves may consist of inverting four, five, or more chips. See what happens in other examples.

85. The Eccentric Exam

Gustavo Piñeiro provides the following explanation: When you answer six questions, the only way to pass the exam is if you have answered all six correctly (one incorrect answer and you do not pass). It is better to answer five and skip the questions you're not sure of. With seven answers, you may have six correct and one incorrect, but there is still the possibility of having an incorrect answer, so it is worth it to try your luck with that last question.

86. Unaligned Pawns

Carlos Serraute was able to fit in 15 pawns.

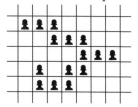

Pablo Haramburu was co-creator of this problem.

87. Square Stamps

With five stamps, we can form 12 small squares. With six stamps, 17; and with seven stamps, 23 small squares.

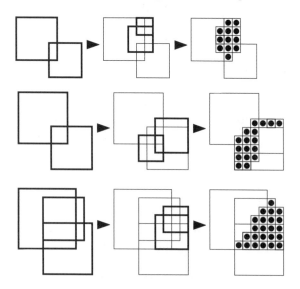

The diagrams show the successive applications of the stamps, which makes it easier to see the impression made by each one. The circles indicate the small squares of size 1 formed in each example.

Without letting the small stamps protrude over the larger, we can form eight small squares with five stamps; ten with six stamps; and fifteen with seven stamps.

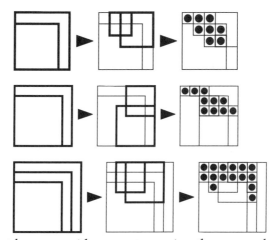

See what happens with more stamps in other examples.

88. The Ghost Detective

14	15	16	19	20
13	12	17	18	1
10	11	6	5	2
9	8	7	4	3

89. The Detective and His Double

5	4	3	2	1
6	**10**	9	8	7
7	8	**9**	10	**6**
1	2	3	4	5

90. Banana Peel

In 10 moves: C right, B left, A down, A right, C up, C left, A up, B down, D left, D up.

91. Fifteen Sweep

1. Four cards are needed: 1, 2, 4, and 7, to force the next player to pick up cards regardless of the card played.

2. The highest number of odd-numbered cards that may be on the table is ten: 5, 5, 7, 7, 7, 7, 9, 9, 9, and 9.

3. A maximum of sixteen cards may be on the table and result in an equal number of even-numbered and odd-numbered cards. For example, four ones, four eights, four nines and four tens.

4. A player could never take just one card (there must be at least two) or 39 (because in this case the opponent would have taken just one, and we've already said this is impossible).

92. Iron Way

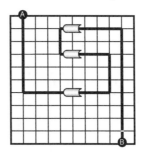

93. The E-Mail Riddle

The messages were sent in this order: E, B, H, C, G, D, A, F.
Ferran Jordá was co-creator of this problem.

GLOSSARY

Figure, Digit, Number

There are ten digits; the numbers 0 to 9. Figures are the graphic symbols representing the numbers. Numbers can have any number of digits.

Divisor, Divisible, Multiplier

Let's take two numbers, A and B. If the result of dividing A by B is exact, with no remainders or decimals, we can say B is a divisor of A. We can also say A is divisible by B. And, if that were not enough, we can say A is a multiple of B.

Number Squared

A number squared is the result of multiplying a whole number by itself. It is sometimes called a perfect square.

The following table shows all numbers squared up to 10,000.

1	4	9	16	25	36	49	64	81
100	121	144	169	196	225	256	289	324
361	400	441	484	529	576	625	676	729
784	841	900	961	1024	1089	1156	1225	1296
1369	1444	1521	1600	1681	1764	1849	1936	2025
2116	2209	2304	2401	2500	2601	2704	2809	2916
3025	3136	3249	3364	3481	3600	3721	3844	3969
4096	4225	4356	4489	4624	4761	4900	5041	5184
5329	5476	5625	5776	5929	6084	6241	6400	6561
6724	6889	7056	7225	7396	7569	7744	7921	8100
8281	8464	8649	8836	9025	9216	9409	9604	9801
10000								

Number Cubed

A number cubed, or sometimes simply a cube, is the result of multiplying a whole number by itself twice.

For example, take the number 3, multiply it by 3, and the result is 9. Then, multiply this result again by 3, which is 27. So, 27 is a number cubed.

Whole Number

Ordinary numbers that are whole numbers, zero, and negative numbers.

Prime Number

A prime number is one that has only two divisors: 1 and itself. Thus, 23 is a prime number, because its only divisors are 1 and 23, whereas 24 is not a prime number, because its divisors are 1, 2, 3, 4, 6, 8, 12, and 24. Take note that according to this definition, 1 is not a prime number.

The following table lists all the prime numbers up to 1013.

2	3	5	7	11	13	17	19	23	29	31
37	41	43	47	53	59	61	67	71	73	79
83	89	97	101	103	107	109	113	127	131	137
139	149	151	157	163	167	173	179	181	191	193
197	199	211	223	227	229	233	239	241	251	257
263	269	271	277	281	283	293	307	311	313	317
331	337	347	349	353	359	367	373	379	383	389
397	401	409	419	421	431	433	439	443	449	457
461	463	467	479	487	491	499	503	509	521	523
541	547	557	563	569	571	477	587	593	599	601
607	613	617	619	631	641	643	647	653	659	661
673	677	683	691	701	709	719	727	733	739	743
751	757	761	769	773	787	797	809	811	821	823
827	829	839	853	857	859	863	877	881	883	887
907	911	919	929	937	941	947	953	967	971	977
983	991	997	1009	1013						

Ordinary Number

Ordinary numbers are those used for everyday counting: 1, 2, 3, 4, 5, 6, and 7, etc. One usually goes to the market to buy 1, 2, or 3 fish, but not 0 fish or −2 fish.

Index

WHAT IS AMERICAN MENSA?

American Mensa
The High IQ Society
One out of 50 people qualifies
for American Mensa ...
Are YOU the One?

American Mensa, Ltd. is an organization for individuals who have one common trait: a score in the top two percent of the population on a standardized intelligence test. Over five million Americans are eligible for membership ... you may be one of them.

• Looking for intellectual stimulation?
You'll find a good "mental workout" in the *Mensa Bulletin,* our national magazine. Voice your opinion in the newsletter published by your local group. And attend activities and gatherings with fascinating programs and engaging conversation.

• Looking for social interaction?
There's something happening on the Mensa calendar almost daily. These range from lectures to game nights to parties. Each year, there are over 40 regional gatherings and the Annual Gathering, where you can meet people, exchange ideas, and make interesting new friends.

• Looking for others who share your special interest?
Whether your interest might be in computer gaming, Monty Python, or scuba, there's probably a Mensa Special

Interest Group (SIG) for you. There are over 150 SIGs, which are started and maintained by members.

So contact us today to receive a free brochure and application.

American Mensa, Ltd.
1229 Corporate Drive West
Arlington, TX 76006
(800) 66-MENSA
Americanmensa@mensa.org
www.us.mensa.org

For Canadians, contact:
Mensa Canada Society
329 March Road
Suite 232, Box 11
Kanata, Ontario Canada
K2K 2E1
(613) 599-5897
info@canada.mensa.org

If you don't live in the U.S. and would like to get in touch with your national Mensa, contact:
Mensa International
15 The Ivories
6-8 Northampton Street, Islington
London N1 2HY England